Colourful Chemistry

F. C. Brown, M.A., B.Sc., F.R.I.C.
Senior Lecturer in Physical Sciences,
University of London Institute of Education

The English Universities Press Ltd
St Paul's House Warwick Lane London EC4

To the user of this book

Chemistry frequently involves colour, and this book tells you how to carry out a number of experiments in which colour plays an essential part. Those of you who are not studying chemistry will find enjoyment in extracting and using dyes, in making pigments, turning them into real paints and then using these paints. Those of you who are studying chemistry will derive added pleasure and profit from understanding the reactions, the tests and the techniques you are employing.

The experiments are neither difficult nor risky. Many of them could be done in the home, and certainly without the need for a proper laboratory. The experiments are designed mainly for C.S.E. and G.C.E. 'O' level students of chemistry and general science.

Acknowledgements

The author gratefully acknowledges the help given by Mr R. Morrison and Mr W. A. Shuttle in trying out and evaluating numerous 'recipes' in the hunt for suitable pigments and dyes. He also thanks most warmly Naomi Coombes for suggesting the idea for the front-cover motif and for demonstrating, in several exciting paintings, that the paints prepared from these instructions can successfully be used by the artist.

ISBN 0 340 11509 2

First printed 1970

Printed and Bound in Great Britain for
The English Universities Press Limited
by Chigwell Press Limited, Albert Road, Buckhurst Hill, Essex

Contents

Section A pH-values

Using pH-paper

All aqueous solutions possess a pH-value of considerable significance to the chemist. In ordinary practice the scale of pH-values lies between the limits of 0 and 14. The value 0 corresponds to extreme acidity while the value 14 corresponds to extreme alkalinity. You should make yourself familiar with this pH-scale. Even though it appears a purely arbitrary scale, there is sound theory behind it.

The simplest way to get used to this concept of pH-value is to use pH-paper, which is paper treated with a carefully blended mixture of indicator-dyes, so that it shows a definite colour when touched with any water-wet material. The colour shown depends on the pH-value and this value can be found by checking the colour against a colour-code usually printed on the container.

Carry out two sets of tests with pH-paper, using quite small pieces of the paper for each test and making sure that your fingers are clean when handling them. For the first set of tests, investigate the pH-value of things *at home*, such as soaps, detergents, toothpaste, vinegar, sweets, fruits, vegetables, sugar, salt, your own saliva, and so on. Remember that everything you test needs to be *wet*. Display your findings on the chart shown below (or on a larger copy of it).

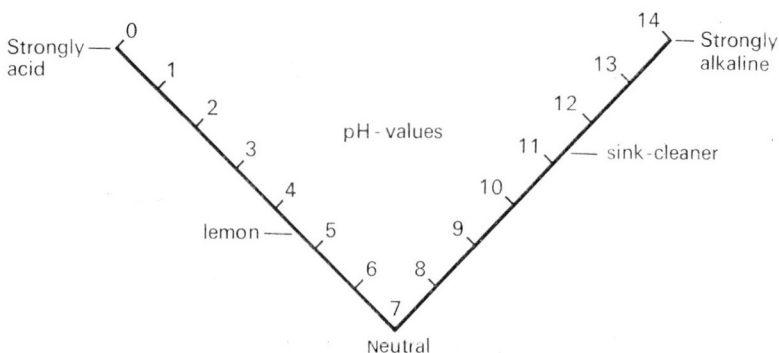

A pH-value chart (to be completed)

For the second set of tests work in the chemistry laboratory, testing drops of solutions or touching the paper on to moistened crystals. Range widely in your tests, and include the following: dilute acids and alkalis, dilute solutions of $CuSO_4$, $NaCl$, Na_2CO_3, KNO_3, NH_4Cl, $Ca(OH)_2$, $MgSO_4$, $FeSO_4$, $ZnSO_4$, $NaOOC.CH_3$.

Again, display your results as you did in the previous set of tests. You will probably find some results rather surprising. Can you explain, for example, the difference in pH-values for NaCl and NH₄Cl, or for NaCl and Na₂CO₃? These pH-values are very reliable and are used by the chemist as an aid to identification. Suppose, for instance, you had two colourless solutions, one of which you knew to be a solution of potassium nitrate and the other a solution of nitric acid, how would you decide which was the potassium nitrate solution?

pH-values and buffer solutions

Now try this experiment. Rinse a beaker with purified (distilled or deionised) water and then run in 10 ml of this water. Check the pH-value of this water. Add from a dropping-pipette or a teat-pipette, *two drops* of M/1 hydrochloric acid. Swirl the liquid to mix. What is the pH-value now?

Carry out a similar experiment in another clean beaker adding, however, *two drops* of M/1 sodium hydroxide solution instead of the hydrochloric acid. What is the final pH-value? Now add the two solutions together in one beaker and find the pH-value of the mixture. This experiment shows how drastically the pH-value of water is changed by very small additions of acid or alkali. Such changes of pH-value could be fatal if they occurred in an animal, such as yourself. What is your pH-value? You found that, approximately, when you tested your saliva. What happens to your pH-value when you eat something acid, like lemon? How does the body keep its pH-value constant, despite the effects of ingested acids and alkalis?

The following experiment provides a clue to the answer to this last question. Dissolve about 0·5 g of sodium hydrogen carbonate, NaHCO₃, in 20 ml of warm water. Cool the solution and find its pH-value. Now add M/10 hydrochloric acid, *drop by drop*, with stirring or swirling, testing the pH-value as you do so. How does this value change? Now start adding drops of M/10 sodium hydroxide solution to the mixture, testing with pH-paper.

This experiment reveals that a solution of sodium hydrogen carbonate is resistant to changes of pH-value when acid or alkali are added to it. Such a solution is called a 'buffer' solution. This one is of special interest to us because sodium hydrogen carbonate does occur in the blood and its presence helps to safeguard the body against 'intruding' acids and alkalis.

Section B Coloured Ions

Coloured ions in solution

You should think of compounds like copper sulphate as orderly arrangements of ions, rather than as 'molecules' of copper sulphate. Copper sulphate is an aggregate of copper cations and sulphate anions regularly arranged in the crystal but disintegrating into a random distribution of these ions when the solid dissolves in water. Colour in such substances, and in their aqueous solutions, usually arises from one of the ions present, but rarely from more than one (copper chromate is an exception—see below). If you consider solutions of, say, potassium sulphate and copper sulphate you will easily deduce that the copper ion, alone, is coloured.

Using ion-exchange resins

It is possible to 'isolate' ions by use of ion-exchange resins. The ions are not actually isolated—that would be impossible—but they can be separated from their opposite numbers (i.e. cations from anions) by becoming temporarily attached to the ion-exchange resins.

For the experiments that follow you will require small quantities (50 g will do) of cation-exchange resin and anion-exchange resin, made by

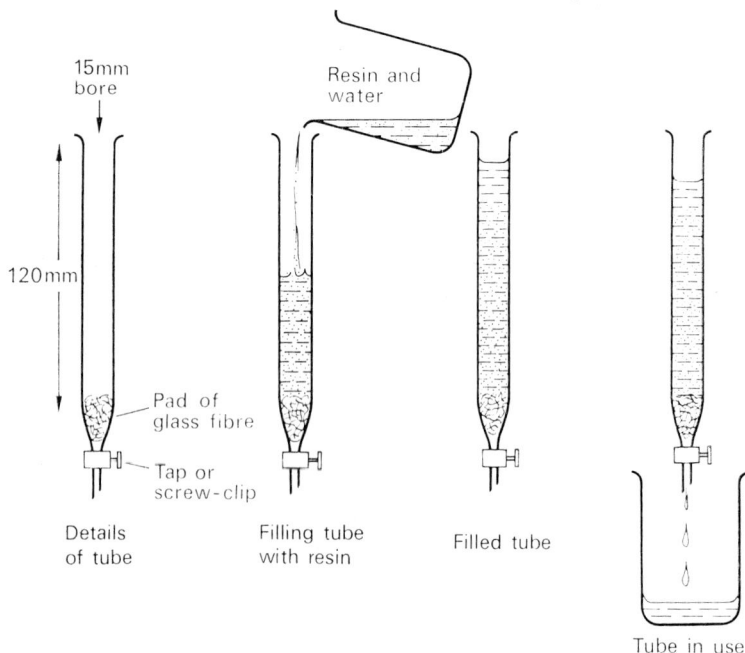

15mm bore

Resin and water

120mm

Pad of glass fibre

Tap or screw-clip

Details of tube

Filling tube with resin

Filled tube

Tube in use

firms such as Permutit, and obtainable from most suppliers of chemicals.

Prepare, as shown in the diagram, two ion-exchange columns, one containing cation-exchange resin and the other anion-exchange resin. The tubes should always be kept filled with water or solution, so that the resins do not dry out. When the tubes are filled with resin, prepare each for action in the following way.

Into the cation-resin tube pour 10 ml of 2M hydrochloric acid, so that the resin is soaked in the acid. Leave for 10 minutes and then wash the tube through with a steady trickle (so that it emerges from the tap below at about 3 drops/second) of purified water until the water emerging, called the effluent, shows a pH-value of 6·5. The tube is now ready.

Treat the anion-resin tube in exactly the same way, but use 2M sodium hydroxide solution instead of the hydrochloric acid. When the effluent shows a pH-value of 7·5, the tube is ready.

Now make up 10 ml of a solution of copper chromate, $CuCrO_4$, heating to aid solution and cooling afterwards. If copper chromate is not available, mix equal volumes of moderately concentrated solutions of potassium chromate and copper sulphate, heating and filtering, if necessary, to obtain a clear solution. The presence of potassium and sulphate ions will not interfere with the run of this experiment. Note the colour of the solution of copper chromate. Pour about 5 ml of it into the cation-resin tube, adjusting the tap so that effluent flows out at 3 drops per second. Catch this effluent in a small beaker and, from time to time, test drops of effluent, as they emerge from the tube, with pH-paper. What change do you observe in this pH-value, and what change do you observe in the resin at the upper end of the tube? What colour is the effluent? Can you account for *all* that you have observed? If you can, then you have explained what has happened in this interesting process.

Now run the effluent, collected in the beaker, through the anion-resin tube, again controlling the flow so that effluent emerges at about 3 drops/second. Check the pH-value of this effluent from time to time and look for any changes in the resin. Note also the colour of the effluent.

Both tubes should now be flushed through liberally with purified water. You will note that the water is unable to remove 'stains' from the resins. These 'stains' are, of course, the trapped copper ions and chromate ions. From your observations of changes in pH-values, what ions do you think the resins gave up in exchange for the ions they now hold? What do you think the effluent from the anion-resin tube is? Evaporate some to dryness in a clean glass basin. Do you get the result you anticipated?

The question now arises—can the copper ions and the chromate ions be dislodged from the resins? Clearly, they are not washed off by water.

Treat the cation-resin with some more of the 2M hydrochloric acid (a powerful source of *hydrogen* ions) for some minutes, and then wash through with purified water, at the usual rate. What effluent appears?

In a similar way, treat the anion-exchange resin with dilute sodium hydroxide solution and then wash well with purified water. The cycle of operations is now complete. The use of ion-exchange resins in industry is of considerable importance. For example, in the production of rare and precious metals like uranium from low-grade ores, the ore is extracted with suitable solvent and the very weak solution is passed through beds of cation-exchange resin. The resin retains the uranium ions and these can later be recovered in a more concentrated form. Another common use of this process is in the removal of calcium and magnesium ions from hard water, in order to soften it.

Your pair of ion-exchange resin tubes should be kept, filled up with water, ready to use for operations similar to those already described. Remember that all solutions must be run through the cation-tube *before* the anion-tube. The reverse order will not do.

You will find it useful and helpful to make a list of the coloured salts you can find, together with the formula of the ion responsible for that colour, in each case, setting it out like this:

SALT	CATION		ANION	
	Formula	Colour	Formula	Colour
Copper (II) chromate				
Potassium permanganate				
Iron (III) chloride				
Potassium chromate				
Nickel acetate				

From a consideration of the results shown in your table, you should be able to deduce the colour, for example, of solutions of potassium sulphate, cobalt acetate and copper chloride.

Section C Dyes

A dye is a colouring-matter, soluble in some solvent like water or an alcohol, that will stain materials such as paper or cloth in a fairly permanent manner. Dyes differ markedly from pigments (see Section E), these latter being *insoluble*.

Natural dyes

Most natural dyes are obtainable from plants and some of them have been known and used for thousands of years—woad for dyeing the skin of warriors; saffron, made from the pollen of a species of Autumn Crocus, for dyeing wool and as a food-colouring; madder, from the root of a plant, for dyeing cloth; turmeric, again from a plant-root, for colouring cloth and also as a food-colouring. Dyes such as these are usually soluble in water, and they cling fairly tenaciously to cloth immersed in such solutions, though the colour tends to fade with washing or exposure to light and air.

Obtain samples of some of these traditional, natural dyes—madder, turmeric, cochineal (almost unique in that it is made from an insect) are suggested—and carry out the following experiments:

Grind up fine, or shred, samples of the dye-materials and gently boil (i.e. simmer) portions with water or, better, a mixture of equal volumes of water and ethanol, in a covered beaker for a minute or two. Decant or filter the hot solution so as to obtain a clear extract of the dye and then use this for trying its effect on small trial-pieces of cloth woven out of cotton, wool or nylon. For each test, immerse the trial-piece for about one minute in a hot solution of the dye, remove and rinse thoroughly under tap-water and then dry in warm air. An interesting additional test is to observe the effect of a drop of dilute acid or alkali on the dried, dyed cloth (drop the acid on one end of the piece and the alkali on the other). In a number of examples, you will find that the dye acts like an 'indicator' in the presence of acid or alkali. This is one of the risks attached to dyes, and one reason for protecting one's clothing against splashes of acid or alkali.

Of course, you may want to try other dyes that you come across in the home. Some inks make excellent dyes when diluted with water. The juice of fruits like blackcurrants might be tried.

A further extension of this work might be to investigate the action of a domestic 'bleach-cleaner' on your dyed materials, after you have finished with them. Dilute a little of the bleach with water in a beaker and immerse the dyed trial-piece in the solution for a few seconds. Remove, wash well in water and dry. Is the dye bleached? Is the cloth affected?

It is strongly recommended that you make a record of your observations in the manner shown in the table below:

NAME OF DYE	COTTON	WOOL	NYLON
Madder Plain			
Acid			
Alkali			
Turmeric Plain			
Acid			
Alkali			

Preparing a modern synthetic dye

Some of the finest dyes are made from chemical substances that are, or were, derived from coal. They are sometimes referred to as 'coal-tar dyes'. The following instructions tell you how to prepare a red dyestuff from two important compounds derived from coal-tar—aniline and resorcinol.

Weigh out 6·5 g of anilinium hydrochloride (aniline hydrochloride) and dissolve it in 15 ml of 4M hydrochloric acid warmed, but not boiled, in a small beaker or a large test-tube. Cool the clear solution obtained and then pour it over small pieces of ice covering the bottom of a 250 ml beaker. Dissolve 3·5 g of sodium nitrite, $NaNO_2$, in 8 ml of water and add this solution, *one drop at a time* (use a teat-pipette, as shown in the diagram, to the mixture in the beaker, swirling the contents gently as you do so. The final product will be a pale yellow solution.

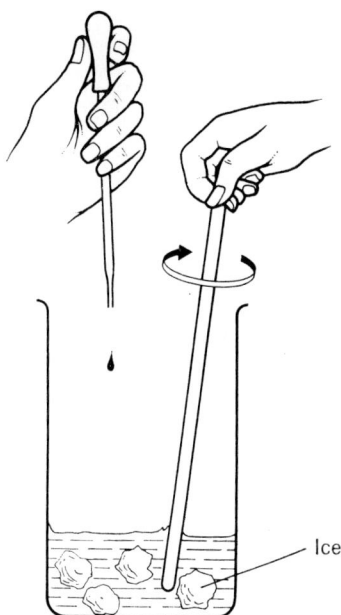

Making a dye

Dissolve 5·5 g of resorcinol in 20 ml of warmed 4M sodium hydroxide solution. Cool this solution before pouring it, very slowly, into the pale yellow solution obtained above, stirring constantly and briskly. The product will be thrown out as a fine red precipitate. Filter and wash well. Set aside part of the precipitate to use in making up one of the paints in Section E (see page 19).

We will call this dye 'Resorcin red'. Boil a little of the product with water to obtain a solution and then use it for dyeing samples of cloth exactly as you did in the first part of the Section. The great advantage of dyes such as this is their resistance to fading. For this reason they are called 'fast' dyes.

Section D Anodising and colouring aluminium

Normally, metal surfaces are painted, but a few metals, notably aluminium, can be coloured most effectively with dyes, giving a permanent and brilliant effect. This can only be done by first oxidising the surface of the metal—a process called 'anodising', and then dyeing this oxide-film. Instructions for carrying this out are given below.

Bend a piece of thin aluminium sheet into a cylinder and fit it as a lining to 400 ml beaker, as shown below. Cut a small rectangle, say 50 mm by 80 mm, of polished aluminium sheet and clean its surfaces by sponging with warm water and detergent, followed by water-rinsing.

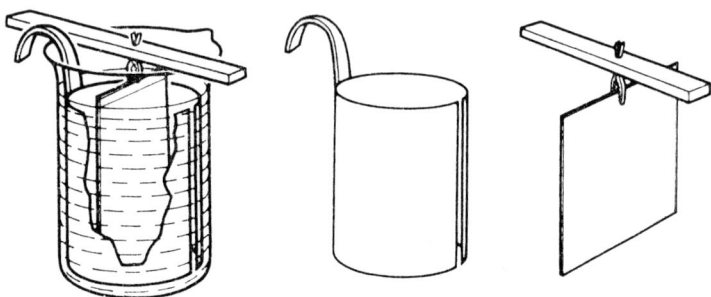

Handle the piece, after washing, with forceps and not with the fingers. Grip it in a 'crocodile-clip' attached to a length of insulated wire and suspend it inside the beaker so that it hangs centrally and not touching the aluminium cylinder. One way of suspension is shown above.

Connect the cylinder to the negative terminal of a low-voltage (6-12 volt) d.c. electrical supply and connect the wire from the central plate to the positive terminal of the d.c. supply (your teacher will check this for you, if necessary). Pour dilute (approx. 2M) sulphuric acid into the beaker until almost all of the suspended piece of aluminium is covered. Then switch on current, starting, if possible, with about 6 volt and slowly raising it to 12 volt. After a few seconds you should see a stream of bubbles rising from the cylinder, and that will indicate that all is going well. Switch off the current after about ten minutes, lift out the aluminium plate, rinse well under the tap and examine it. You will note a distinct difference between the part of the plate that was immersed and the part that was not. This is due to the formation of an almost invisible film of transparent oxide.

You can now dye this oxide-film by dipping the plate into a nearly boiling solution of a dye in water. There are several common dyes to try —eosin, alizarin, purple or green 'Quink', cochineal—and quite a dilute solution will do. Take the aluminium out, from time to time, to decide when the tint is adequate. Then immerse the dyed plate in boiling water in a beaker for a minute or two. This will seal in the dye, to give a permanent effect. If you wish to obtain a gold effect, such as is used on imitation gold jewelry, use as 'dye' a few crystals of ferric chloride and an equal weight of ammonium oxalate (*poisonous—so rinse your hands well*).

Note the brilliant colour of the finished specimens. This is chiefly due to the metal under the dyed oxide film reflecting light up through it. It is worth noting that this oxide film is a very good electrical insulator.

Section E Pigments and paints

Introduction to pigments

A paint is made by grinding a pigment, or mixture of pigments, to a very fine powder and then suspending (not dissolving—note the difference) the powder in a suitable liquid. In ancient times the range of pigments available was very limited, being largely restricted to minerals that occur naturally. Pigments such as vermilion, umbers, siennas, ultramarine and malachite were processed, finely ground and then made up into paints.

Nowadays the skills of the chemist are employed in producing a wide range of pigments, some of them by modernising versions of old processes, some by entirely new methods. In the text that follows you will find directions for making, for your own use, a variety of pigments, using quite simple chemical reactions and techniques. Nearly all the chemicals required are to be found in a well-stocked school chemistry department. You can then prepare paints from these pigments.

Thus, in producing a range of useful 'poster paints', you will carry out interesting chemical reactions and processes and learn some useful techniques. In order to make your work as enjoyable and successful as possible, read and take careful note of the following.

Useful hints

1. All weights are given to the nearest gramme, no greater accuracy being required in such work. Weigh out solids on clean rough paper. Learn to weigh with speed and accuracy.
2. The word 'water' in the instructions means 'purified water'. Clean tap-water can be used instead, but purified (i.e. distilled or deionised) water will give superior results.
3. Work cleanly and tidily. All apparatus should be washed immediately after use and, again, just before use. Rinse your hands frequently. The chemicals you are handling are not dangerous, but *all* chemicals should be handled as though they are toxic if taken into the mouth, and every care must be taken to avoid such a risk.
4. Solids dissolve in water much more readily when they are finely-powdered and when the water is hot. Make use of this advice in making up your solutions, but remember that, unless otherwise instructed, the final solution must be cooled down to room temperature before use.
5. When you are about to mix two solutions to form a precipitate, it is a very good idea to mix single drops of the solutions on a white or

black tile, examining closely to see what actually happens when the two solutions meet (see diagram below). Note that 'add A to B' does not mean 'add B to A'; i.e. instructions must be followed with care.

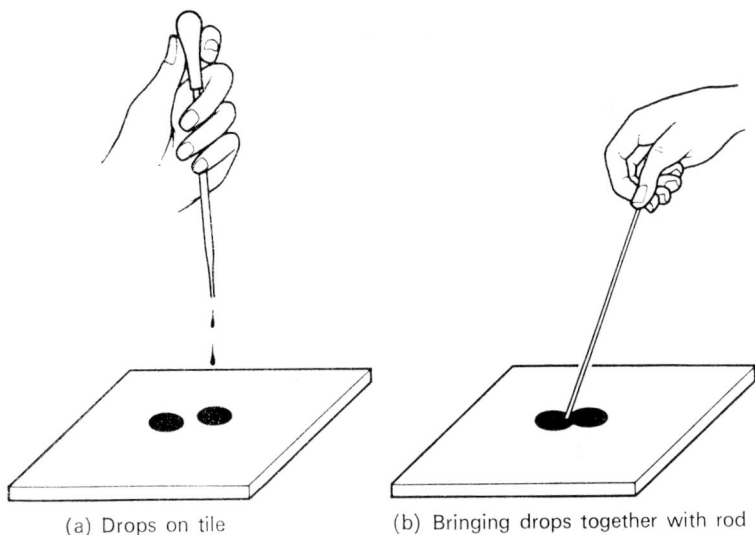

(a) Drops on tile   (b) Bringing drops together with rod

"Trial run"

6. Filtration is used to separate precipitate from liquid. Ordinary filtration-methods will prove slow,in the work that follows, and you are advised to use 'vacuum-filtration' where possible. Most chemistry laboratories have apparatus for doing this. Alternatively, precipitates can be separated by allowing them to settle and then pouring off the top clear liquid (see diagram on p. 16).
Both methods of isolating precipitates are fully described in the first experiment (for making white pigment).

7. To dry a wet precipitate, spread it out on absorbent paper (paper towels are good) and expose to a draught of warm air. Do not heat the precipitate, or put it in an oven to dry—it may discolour or decompose.

8. When the precipitate is very fine, it need not be thoroughly dried before making into paint, and will need no grinding. The instructions will guide you about this. Grinding a pigment is done in a mortar with a pestle. Choose a large, rather than a small pair and grind with a rotary motion of the pestle until the powder is as fine as possible.

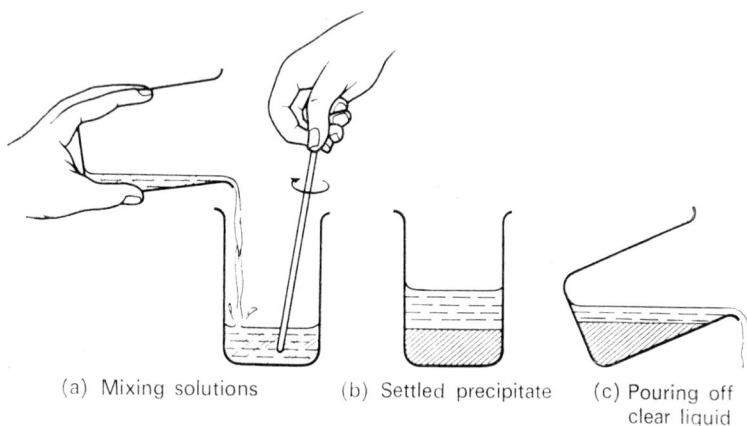

(a) Mixing solutions   (b) Settled precipitate   (c) Pouring off clear liquid

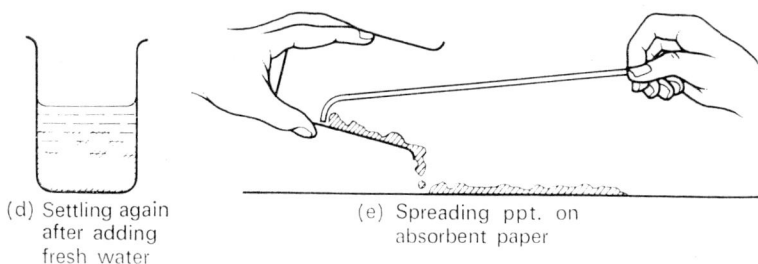

(d) Settling again after adding fresh water   (e) Spreading ppt. on absorbent paper

9. When you make the pigment into the paint, mix it into a smooth paste with added drops of liquid medium. A thin glass rod, or a stout aluminium wire, make good stirrers. A recommended 'recipe' for a good medium is: mix 30 ml of clear gum (not paste—clear office-type gum) with 10 ml of water and one drop of a liquid detergent. Make up plenty of this medium; it keeps quite well. The pasty paint

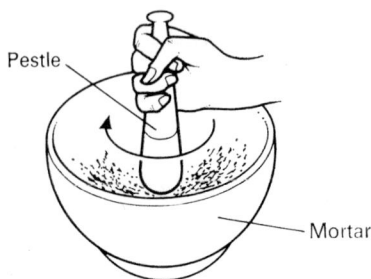

Pestle

Mortar

Grinding a solid

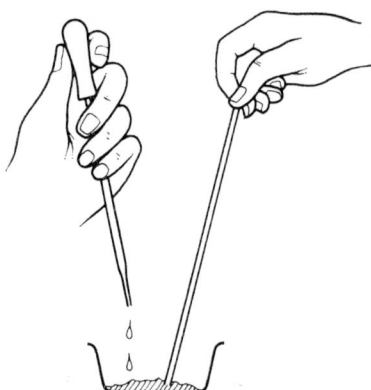
Mixing pigment
with medium

may harden in time but can be softened again by adding a little lukewarm water, which will be absorbed by the paint, restoring it to condition.

10. The more intensely coloured paints, such as Brilliant red or Oxford blue, can be toned down by mixing with white, giving 'pastel' shades.

Preparing paints and pigments

1. White

Paint sample here

Make plenty of the white pigment because you will be using it, as well as in its own right, for mixing with other pigments to get pastel shades.

Weigh out 10 g of anhydrous sodium carbonate, Na_2CO_3, or 28 g of hydrated sodium carbonate. Na_2CO_3 $10H_2O$, and dissolve in 100 ml of warm water. Cool the solution to air temperature. Weigh out 30 g of lead nitrate, $Pb(NO_3)_2$, and dissolve in 150 ml of water. Using a sufficiently large beaker for the purpose, pour the lead nitrate solution slowly, with steady stirring, into the sodium carbonate solution. Wash and separate the white precipitate (which is the required pigment) by one of the two processes described below:

(a) Fit together a clean Buchner funnel and a clean filter-flask. Attach the unit to a suction-pump (usually worked off the cold-water tap). Lay a disc of filter-paper on the base of the funnel, wet it with water and apply gentle suction. Pour the contents of the beaker into the funnel, increasing suction as required. Occasionally, the filtrate comes through cloudy at first. If this happens, detach the flask from the pump, empty it back into the beaker, reconnect to the pump and

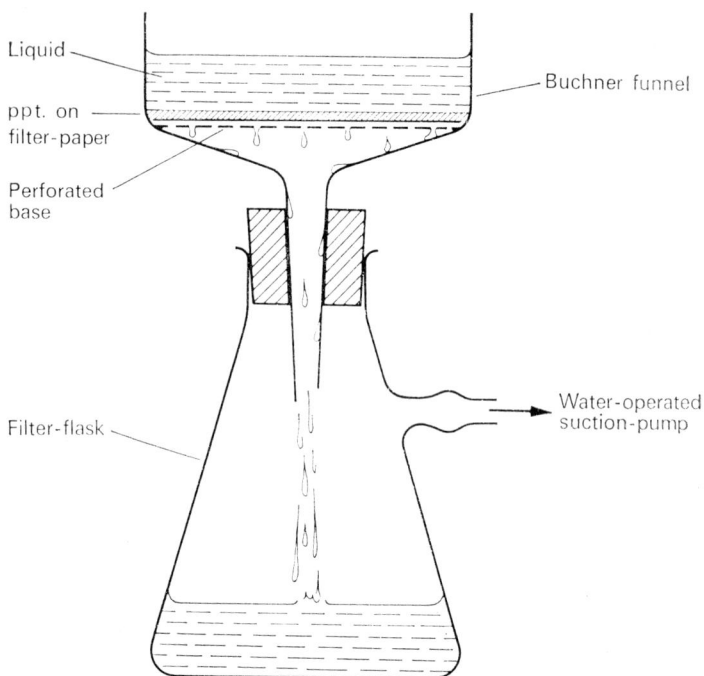

Liquid

ppt. on filter-paper

Perforated base

Filter-flask

Buchner funnel

Water-operated suction-pump

"Vacuum" – filtration

continue filtration. When all liquid has been sucked through into the flask, wash the precipitate by several successive small additions of water, sucking dry each time. Finally, detach from the pump, ease out the cake of damp pigment from the funnel and spread it out on absorbent paper, so that it can be exposed to a draught of warm air, to dry.

(b) If you have no suction-filtration facilities available, allow the contents of the beaker to settle. Carefully pour away the clear, top layer of liquid, add more water, stir well, allow to settle again and pour away the top, clear liquid. Repeat this process several times (three or four at least) before spreading the wet pigment out on absorbent paper to dry.

If you are studying chemistry, you should have no difficulty in understanding the reaction that occurred when you mixed the two solutions, nor in writing down an equation for that reaction.

This pigment needs no grinding and it can be made up into white paint in the manner described in 'Useful hints', page 16.

2. White (alternative)

Paint sample here

Titanium oxide, TiO_2, finely ground, can be mixed with the medium to make an excellent white paint. Take 7 g of the oxide.

3. Black

Paint sample here

Weigh out 3 g of vegetable carbon and stir it with added drops of medium to produce a smooth paste.

4. Yellow

Paint sample here

Dissolve 7 g of potassium chromate, K_2CrO_4, in 50 ml of water. Dissolve 10 g of lead nitrate, $Pb(NO_3)_2$, in about 100 ml of water. If heating is used to aid solution, cool the solutions right down to air temperature before the next stage. Pour the chromate solution, with stirring, into the lead nitrate solution. Treat the fine yellow precipitate, the required pigment, exactly as you did the white pigment in No. 1.

Assuming this product to be lead chromate, try to write down an equation for the reaction.

5. Orange

Paint sample here

Weigh out 10 g of red lead, Pb_3O_4, and make up into paint with the medium.

6. Brilliant red

Paint sample here

There seems to be no way of preparing, in the school laboratory, using simple and safe chemicals, a good red pigment, so it is proposed to make use of a dye, commonly found in the laboratory (and, indeed, in red ink), called eosin, mixing the dye with white lead to produce a composite.

Weigh out 1 g of eosin powder and 2 g of white lead. Mix these with drops of medium to produce a paste. By varying the proportions of eosin and white lead, the shade of red can be adjusted. This particular paint needs very thorough mixing prior to use.

7. Resorcin red

Paint sample here

This paint is, to some degree, an alternative to Brilliant red, though it lacks the brilliance of that paint. It can be prepared only if you have made a supply of the dyestuff according to the instructions on page 10, in the section on 'Dyes'. Mix 1 g of the powdered dye with medium, in the usual way.

8. Dark red

Paint sample here

The pigment for this paint is copper(I) oxide, Cu_2O, which you can prepare by reducing Fehling's solution with glucose.

Weigh out 20 g of copper sulphate crystals and dissolve in 150 ml of water. Weigh out 35 g of Rochelle salt, sodium potassium tartrate, and

17 g of potassium hydroxide (or 12 g of sodium hydroxide) pellets and dissolve these two reagents in 100 ml of water in a beaker. In a sufficiently large beaker mix the two solutions prepared above, giving a fine deep blue solution. This is Fehling's solution. Warm it over a bunsen flame until it is hot, but not boiling—a temperature of 70°C is admirable—and stir in 10 g of powdered glucose. The colour of the solution will slowly change, through dark blue, green, brown, until a bright red precipitate will be thrown out. Allow the contents of the beaker to remain hot for some minutes before filtering or decanting. The precipitate, when washed, need not be completely dried before making up into paint.

9. Red-Brown

| Paint sample here |

Make up 7 g of finely-powdered iron (III) oxide, Fe_2O_3, into paint with medium. This pigment is one of the most ancient known to man and was used by the earliest painters.

10. Dark brown

| Paint sample here |

Make up 10 g of lead (IV) oxide, PbO_2, into paint, with medium.

11. Light green

| Paint sample here |

Dissolve 15 g of copper sulphate crystals in 100 ml of water. Dissolve 5 g of sodium hydrogen carbonate, $NaHCO_3$, in 50 ml of water. Add the copper sulphate solution, with brisk stirring, to the carbonate solution. You will observe the evolution of bubbles of gas (what gas?), as well as the formation of a pale green precipitate. Wash and isolate the pigment and make it up, when roughly dry, into paint in the usual way.

12. Cambridge blue

| Paint sample here |

Proceed exactly as in No. 11, but with the addition of one or two ml of sodium hydroxide solution to the solution of sodium hydrogen carbonate, when making it. The more sodium hydroxide added, the bluer will be the pigment, since $Cu(OH)_2$ is blue in colour.

In this experiment, two reactions are occurring together, and you should be able to write down equations for each of them.

13. Dark green

| Paint sample here |

The pigment for this paint is chromium(III) oxide, Cr_2O_3, and the preparation of the oxide from ammonium dichromate is quite an exciting experiment.

Find a tall 250 ml beaker, preferably with a lip. Stand it on an asbestos-coated wire gauze, on a tripod over a bunsen burner, and have ready a clock-glass big enough to cover the mouth of the beaker completely.

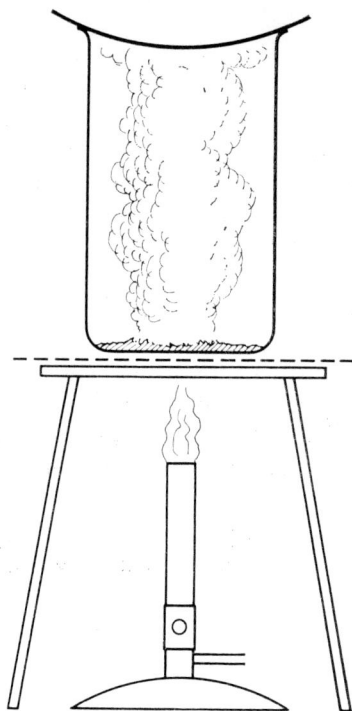

Gently heating
ammonium dichromate

Weigh out 3 g of crystalline ammonium dichromate, $(NH_4)_2Cr_2O_7$, into the beaker, put on the clock-glass cover and apply *very gentle* heat with a *small* bunsen flame under the gauze. In a few seconds the crystals will darken and then a very energetic reaction will occur, throwing up a cloud of green particles. This is the required pigment. Allow it to cool and then grind it up as finely as possible in a mortar before making into paint. The chemistry of this reaction is very interesting. It is a thermal decomposition, and the only products, beside the oxide, Cr_2O_3, are water (steam) and nitrogen. From that information you should be able to work out an equation for the reaction which is, by the way, almost unique in chemistry and is an excellent method for producing a sample of pure nitrogen.

Note. If you require more of this pigment, repeat the above process. Do not try to make larger quantities by heating more than 3 g of ammonium dichromate in one operation; it is somewhat risky and you are likely, in any case, to lose much of the product out of the beaker.

14. Oxford blue Paint sample here

This pigment is a 'modern' one, being made by synthetic processes similar to those used in making the famous pigment, Monastral Blue.

Place in a mortar 6 g of phthalic anhydride, 8 g of urea, 2 g of hydrated copper(II) chloride and a small crystal of ammonium or sodium molybdate (catalyst). Grind these ingredients to a coarse powder and introduce the mixture (or part of it) into a large test-tube or a dish made of Pyrex glass or porcelain. Mount the tube or basin high (20 cm) above a small bunsen flame, so that the mixture is heated very gently. Heating must not be excessive or hurried—the slower you take this reaction the better will be the product.

The mixture will melt to a green liquid, evolve gases, slowly turn dark green and then blue, partly solidifying at this stage. Continue the gentle heating until the fine blue solid first observed has changed to a brownish-green solid. Allow the product to cool, scrape it out of the tube or dish and crush it to a coarse powder. This is the crude product. It must now be purified. Into a 250 ml beaker pour 75 ml of moderately dilute (3 molar) sulphuric acid, add the powdered pigment and warm *gently*, with stirring.

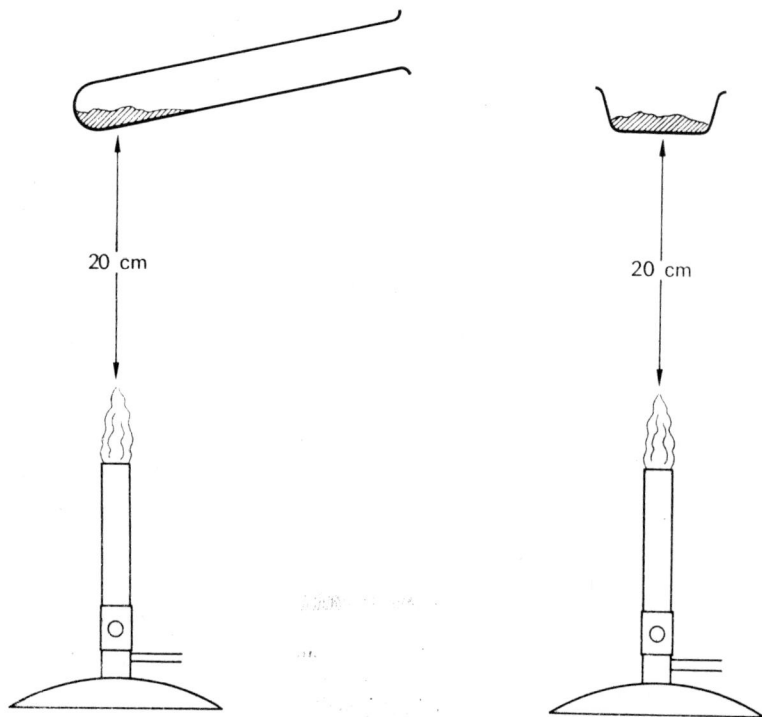

20 cm

20 cm

The blue pigment will appear. Allow the liquid to cool right down to air temperature and then pour it into a larger volume of water in a second beaker. The precipitate can now be filtered or isolated in the usual way. It should then be thoroughly dried and finely ground before making up into paint. The purified product is a very intense and economical pigment, so do not be concerned if the yield appears small.

| Paint sample here |

15. *Prussian blue*

Dissolve 5 g of hydrated iron (III) chloride, $FeCl_3 6H_2O$, in 50 ml of water. Dissolve 10 g of potassium ferrocyanide, $K_4Fe(CN)_6 3H_2O$, in 75 ml of water. Add the first solution to the second, stirring briskly. The dark blue precipitate will prove difficult to wash and filter but, with patience, you will be able to reduce it to a wet mass that will slowly dry off in warm air. There is no need to dry the pigment completely before making it into paint in the usual way.

| Paint sample here |

16. *Gold*

Mix the desired quantity of copper ('bronze') powder with medium.

17. *Silver*

| Paint sample here |

Mix 'Flake aluminium powder' with medium to a thick paste.

If you have carried out all, or most of the preparations in this Section, you will now have an excellent range of pigments and paints with which, no doubt, you will want to produce some paintings. There is no reason why you should stop at that. Other pigments can be prepared, and you may wish to experiment further to see whether you can find new or better ones. If you do so, note the following points:

(a) a pigment must be stable in air,

(b) a pigment should be insoluble in water. The one called Brilliant red, in this section, is not a true pigment, because it is partly made of soluble dye.

(c) when trying out a new reaction, use very small quantities at first (see Hint No. 5 on page 14).

(d) salts of metals like cobalt, nickel, chromium, iron, copper are 'likely starters' for preparing pigments.

Here are two problems that you might like to look into:

(a) Can you make a true pigment that is bright red?

(b) Can you make a primrose-yellow pigment?

PIGMENT and PAINTS: a SUMMARY

Description	Pigment used	Reagents required, and other information
White	'White lead'	$Pb(NO_3)_2$ and Na_2CO_3. Contains $PbCO_3$ and $Pb(OH)_2$
White (alternative)	Titanium dioxide	Used as purchased
Black	Vegetable carbon	Used as purchased
Yellow	Lead chromate	$Pb(NO_3)_2$ and K_2CrO_4. Known as 'chrome yellow'
Orange	'Red lead'	Finely ground—chiefly Pb_3O_4
Brilliant red	Eosin and white lead	Eosin is not a pigment, but a dye
Resorcin red	See page 10	This is a dyestuff used as a pigment
Dark red	Copper(I) oxide	Fehling's solution and glucose
Red-brown	Iron(III) oxide	Fe_2O_3, finely ground
Dark brown	Lead(IV) oxide)	PbO_2, finely ground
Light green	Copper carbonate	$CuSO_4$ and $NaHCO_3$. Pigment is largely $CuCO_3$
Cambridge blue	Basic copper carbonate	Contains both $CuCO_3$ and $Cu(OH)_2$
Dark green	Chromium (III) oxide	$(NH_4)_2Cr_2O_7$, thermally decomposed
Oxford blue	Copper phthalocyanine	Phthalic anhydride, urea and $CuCl_2$
Prussian blue	Approx. formula $Fe_7(CN)_{18}.9H_2O$	$FeCl_3$ and $K_4Fe(CN)_6$
Gold	Copper-bronze powder	Used as purchased
Silver	Flake aluminium powder	Used as purchased

PUBLISHER'S NOTE

The apparatus and materials required for the experiments described in this book are commonly available in the school chemistry laboratory. Messrs. Griffin and George, however, market a kit specially designed to meet these requirements for a small group of students. The kit contains certain special items of equipment and almost all the chemicals (dilute acid and alkali excepted) used in the experiments.